Simeon Hayden Guilford

Nitrous Oxide

Its Properties, Method of Administration and Effects

Simeon Hayden Guilford

Nitrous Oxide

Its Properties, Method of Administration and Effects

ISBN/EAN: 9783337345730

Printed in Europe, USA, Canada, Australia, Japan

Cover: Foto ©berggeist007 / pixelio.de

More available books at **www.hansebooks.com**

NITROUS OXIDE;

ITS PROPERTIES, METHOD OF ADMINISTRATION AND EFFECTS.

BY

S. H. GUILFORD, A. M., D. D. S.,

PROFESSOR OF OPERATIVE AND PROSTHETIC DENTISTRY AT THE PHILADELPHIA DENTAL COLLEGE.

PHILADELPHIA:
SPANGLER & DAVIS, PRINTERS.
1887.

Entered according to Act of Congress in the year 1887, by
S. H. GUILFORD,
in the office of the Librarian of Congress at Washington, D. C.

"The knife of the surgeon is steeped in the waters of forgetfulness, and the deepest furrow in the knotted brow of agony is forever smoothed away."

O. W. HOLMES.

"If America had contributed nothing more to the stock of human happiness than anæsthetics, the world would owe her an everlasting debt of gratitude."

S. D. GROSS.

PREFACE.

THE preparation of this Manual was prompted by the feeling, expressed by others and felt by the writer, that such a work would prove a benefit to the student and young practitioner. This feeling was emphasized by the fact that the last practical work upon the subject, published in this country, has been out of print for sixteen years.

The record that nitrous oxide has made for itself since its introduction proves it to be the safest of all anæsthetics, and the one most suitable for minor operations in dental surgery.

Its employment is being extended so rapidly that a better knowledge of its properties and effects seemed necessary to enable novices to secure its full advantages without incurring undue risk of accident. With this end in view the author has endeavored to treat the subject in a thoroughly practical manner, giving full details of preparation, administration and extraction under its influence, and at the same time indicating the manner of dealing with any emergencies that might arise in connection with its employment.

The last two chapters, treating of the combination of anæsthetic agents and of legal questions related to the administration of anæsthetics, will, it is hoped, prove to be of special value to those for whom the work has been prepared.

The author would acknowledge the kind assistance of Dr. E. C. KIRK in the revision of proof, and the courtesy of publishers and dealers in furnishing certain cuts, which are credited to them in place.

S. H. G

CONTENTS.

Chapter.		Page
1.	HISTORY,	1
2.	CHEMICAL PROPERTIES,	5
3.	PHYSIOLOGICAL ACTION,	8
4.	RELATIVE SAFETY,	22
5.	ADVISABILITY OF ADMINISTRATION IN SPECIAL CASES,	26
6.	MANUFACTURE,	32
7.	INHALERS AND ACCESSORY APPLIANCES,	45
8.	ADMINISTRATION,	52
9.	EXTRACTION DURING ANÆSTHESIA,	61
10.	ACCIDENTS AND EMERGENCIES,	70
11.	COMBINED ANÆSTHETICS,	79
12.	LEGAL CONSIDERATIONS,	87

NITROUS OXIDE.

CHAPTER I.

HISTORY.

The discovery of nitrous oxide, or nitrogen protoxide, with its valuable properties and resultant benefits to mankind, was not the work of one man or of one time. Like many other valuable discoveries and inventions, the credit of its origin and successful use is divided among several, each of whom contributed his part toward the final result.

Thus the discovery of the gas, *per se*, was made by Priestley in 1776; its exhilarating and analgesic effects were noticed by Davy in 1800; while its full and practical value as a true anæsthetic was first demonstrated by Dr. Horace Wells, in 1844. Davy suggested that "as nitrous oxide, in its extensive operations, appears capable of destroying physical pain, it may probably be used with advantage during surgical operations in which no great effusion of blood takes place." From 1800, when the above prophetic suggestion was made, up to the time of its introduction as an anæsthetic, in 1844, its anæsthetic properties received no attention and called forth no experimentation of which we have any knowledge on the part of the medical or dental professions.

Its exhilarant properties, however, were at once taken advantage of and demonstrated both in the class-room and on the lecture platform.

Both abroad and in this country during those forty-four years lecturers appeared from time to time in various places discoursing upon the peculiar properties of this "laughing gas" and exhibiting its effects upon subjects, to the amusement of the audience and the pecuniary benefit of the exhibitor.

Mr. G. Q. Colton, of New York, one of the more prominent lecturers, gave an entertainment of this kind at Hartford, Conn., in December, 1844. Dr. Wells, a dentist of that city, was present as a spectator, and noticed the freedom from pain that attended the accidental injury of one of the subjects. This fact deeply impressed Dr. Wells, who was led to believe, from what he had seen, that teeth might be extracted without pain under its influence.

Determined to test the matter upon his own person by having a tooth extracted, he arranged with Mr. Colton to come to his office on the following day to administer the gas, while Dr. John M. Riggs, another dentist of the same city, was invited to be present and perform the operation. Dr. Wells inhaled the gas at the hands of Mr. Colton, and Dr. Riggs extracted the tooth. On recovering, he declared that lack of consciousness and absence of pain were both complete, and then and there prophesied a great future for the new and only true anæsthetic.

Dr. Wells immediately communicated his discovery to others, and began, in conjunction with Dr. Riggs, an extended series of experiments with the new agent. He used it freely during the whole remaining period of his dental practice, and other dentists, particularly in the Eastern States, adopted its use.

Eminently satisfactory as it had proven in dental operations, Dr. Wells was not content to allow its use to rest there. He believed that there was a great field for it in general surgery, and began urging its introduction in that direction. In all minor operations, and some major ones, it answered well, as it had in those pertaining to the mouth. There were, however, two serious objections to its use in general surgery. One was the care and knowledge necessary for its proper preparation, and the other its bulkiness, rendering it difficult to transport from place to place.

These objections led Dr. Wells to cast about for some other anæsthetic that would be free from them. Sulphuric ether was known to possess certain anæsthetizing properties, and Dr. Wells at once began to experiment with it. His efforts and experiments to prove its value as a practical anæsthetic in 1845 were quite successful, and others then taking it up, it soon came into general use in surgery, and has so continued ever since.

It will thus be seen that for the practical introduction of both nitrous oxide gas and sulphuric ether as anæsthetic agents the world in general is laid under obligations to Dr. Horace Wells, an American dentist. He died in 1848, too early to receive the honors and blessings of grateful humanity, but with his name will ever be linked the grandeur of his achievements.

From his day to ours nitrous oxide has been the adopted anæsthetic of the dental profession. The duration of its effects, though brief, is sufficient for our purposes, while its comparative freedom from danger makes its use desirable for all minor operations.

Its employment became more general year by year, and its extension would have been more rapid but for the difficulty attending its manufacture. In the earlier days it was made by the individual using it, which required a rather expensive and cumbrous apparatus, besides demanding a fair amount of chemical knowledge to ensure its freedom from deleterious contaminations. Within the past dozen years, however, these objections to its more general use have been overcome by its careful manufacture on a large scale by experienced and skillful hands and its compression into liquid form in strong iron bottles or cylinders. Its compactness and convenience, together with its absolute purity in this form, have greatly extended its use, and must continue so to do for all dental and many surgical purposes.

CHAPTER II.

CHEMICAL PROPERTIES.

Nitrous oxide is a transparent, colorless gas, of slightly sweetish taste, and faint but pleasant odor. Its chemical symbol is $N_2 O$, indicating that it is composed of two equivalents by volume of nitrogen to one of oxygen. The symbol was formerly written $N O$, because it was known that in this gas nitrogen and oxygen were combined in their lowest combining quantities, but as oxygen is bivalent and nitrogen in this compound is univalent, it is properly written $N_2 O$. Although known as nitrous oxide, it is in reality hypo-nitrous oxide, nitrous oxide proper being represented by the symbol $N_2 O_3$. Its specific gravity is 1.527, one hundred cubic inches weighing 47.29 grains.

Under a pressure of 50 atmospheres at 45° F. it is condensed into a clear, transparent liquid, and at 150° F. below zero this liquid solidifies. Nitrous oxide is not a simple mixture of its two constituent gases, but a definite chemical compound in which the properties of the original gases are lost and new ones peculiar to itself are developed. Proof of this is found in the fact that the gas has both a taste and a smell, while air, composed of the same elements, has neither. Thus, also, hydrogen and oxygen, when simply mixed, can be inhaled for a time without ill effects, while water, a chemical compound of the same elements, is totally irrespirable. Nitrous oxide will also support combustion, provided the substance introduced into it be first heated sufficiently to resolve the compound into its original elements.

It may be produced from various substances, but is most conveniently and profitably manufactured from ammonium nitrate. This salt, when subjected to heat, first melts, then boils and undergoes decomposition, the gas being liberated at about 400° F. In its decomposition only nitrous oxide and water are formed, thus:

$$NH_4NO_3 + \text{heat} = 2H_2O + N_2O.$$

Impurities are sometimes found in the commercial salt, the most common and dangerous one being chlorine. As this is somewhat difficult of separation from the nitrous oxide, it is better to see in advance that it is not present in the salt. Ammonium nitrate is readily soluble in five or six times its bulk of water. If such a solution be made, and a solution of argentic nitrate be added, the presence of chlorine will be indicated by the mixture becoming clouded, due to the formation of argentic chloride. Should the solution remain clear we may be sure that chlorine is not present. All new lots of the salt should be thus tested before using.

CHEMICAL PROPERTIES.

Nitrous oxide is somewhat soluble in water, the latter taking up its own bulk or more of the former. It may again be eliminated by subjecting the water to a moderate degree of heat. Besides combining to form nitrous oxide, the two gases, nitrogen and oxygen, unite in different proportions, forming a series of chemical compounds as follows : *

N_2O, Hypo-nitrous oxide.
N_2O_2, Nitrogen di-oxide.
N_2O_3, Nitrous oxide.
N_2O_4, Nitrogen tetra-oxide.
N_2O_5, Nitric oxide.

Although nitrous oxide, in composition, differs from all other chemical bodies, it is somewhat analogous to atmospheric air. They have the same constituent elements, both are supporters of combustion under certain conditions, and both may be respired in proper quantities without harm. They differ from each other, however, both in the proportion of their elements and in the character of their association. Thus, while nitrous oxide contains one-third of oxygen to two-thirds of nitrogen, atmospheric air is composed of one-fifth of oxygen to four-fifths of nitrogen. In addition to this, as previously stated, in nitrous oxide the elements are in chemical combination, whereas in air they are simply associated mechanically.

Nitrous oxide is more respirable than oxygen and possesses the property of anæsthetization, a quality totally lacking in simple oxygen.

---

* Barker's Chemistry.

CHAPTER III.

PHYSIOLOGICAL ACTION.

The exact manner in which nitrous oxide acts upon the system to produce its well-known effects has never been clearly demonstrated. Like its principal rivals, sulphuric ether and chloroform, a great variety of opinions has always existed in the minds of the various investigators as to how and why these effects are produced.

The outward manifestation of their action is, in the case of each one, well marked and clearly defined, and from these manifestations we know to a great extent what organs are affected; but the precise manner in which these organs are operated upon by the anæsthetic agent must be left for the future to determine.

All of these three most commonly used agents have many points of similarity in their action, but they differ widely both as to the manner in which they affect the various organs and the order in which these organs become affected in the course of anæsthetization. Thus, each one, when administered in quantity and manner consistent with its nature and the condition of the patient, will produce the same effect—anæsthesia; and each one, also, when the administration is carried beyond this point, may or will produce death.

So, too, we know that the action of the lungs, heart and brain are modified by them through the agency of the circulating medium, the blood; but each one of these organs is affected differently, in degree, in order, and probably in manner, by the different agents employed.

The external symptoms, as they manifest themselves, indicate that the individual passes through three stages in the process of anæsthetization. First, we have *exhilaration*, or stimulation, then *excitement*, and afterwards *relaxation*, accompanied by suspension of sensation and motion.

These stages are very well marked in the administration of ether, while they are less so with chloroform and nitrous oxide.

In this fact lies to a great extent the comparative safety of ether, since it enables the operator to readily determine how far the patient is being carried, and to cease the administration at the proper moment. Chloroform is more dangerous, because with it these symptoms are often confluent or follow each other so rapidly that the limit of safety may readily be passed without proper warning.

Nitrous oxide differs from all other anæsthetics in several important particulars, which are worthy of being mentioned in some detail. First of all, it is agreeable to the individual while being inhaled, for, aside from its slightly sweetish taste, the patient would not be able to distinguish it from atmospheric air. Both ether and chloroform in the first stage of administration produce a marked irritating effect upon the respiratory track that is both unpleasant and alarming to the patient, since it produces a feeling akin to suffocation. No such feeling is produced by nitrous oxide; hence, we are enabled to exhibit it in full quantity from the very beginning without admixture of air, a very important matter in the production of narcosis with this agent.

Another important property of nitrous oxide is the rapidity with which it produces its effects. With a more dangerous agent this would not be desirable, owing to the attendant risk of passing the limit of safety without proper warning; but, with so safe an agent as this, such risk is reduced to its smallest proportions. Besides this, abundant experience has proven that the best results are obtained from nitrous oxide when the patient is under its influence for the shortest time. This is probably due to the fact that in such case the oxygenation of the blood is interfered with for a shorter period.

By the administration of nitrous oxide in full quantity the average patient is fully anæsthetized in thirty seconds. This condition continues complete for about a minute after the administration is suspended, after which about another half minute elapses before the patient is fully restored to consciousnesss. The whole time thus occupied from end to end for any dental operation does not usually exceed two minutes. The rapid elimination of the gas from the system during recovery is a feature peculiar to itself and greatly appreciated by both patient and operator, for, beside the saving of time to both, it relieves the latter from the nervous strain always existing when his patient is in an unconscious condition.

Another peculiarity in the action of nitrous oxide is that when anæsthetization is complete, sensation is completely suspended, but consciousness may not be entirely so, and while there is suspension of muscular action, there is no general muscular relaxation.

With all true anæsthetics, while their influence is complete, there must, of necessity, be absolute suspension of sensation; but with most of them there is also complete suspension of consciousness attended with absolute muscular relaxation.

A patient fully under the influence of ether or chloroform, besides being insensible and totally unconscious, has his muscular system so thoroughly relaxed that it is often difficult to keep the body upright in the chair, and the arm, if raised, will fall limp by the side. Under the influence of the gas, however, the patient easily maintains an erect position, and the arm, if lifted and released, will either remain upright or descend slowly to its former position. So, too, patients, on recovering from the effects of the gas, will often be able to tell how many teeth were extracted and the order in which they were removed, although they had been entirely oblivious to any sensation of pain. They sometimes also speak of a desire to put forth muscular effort while under the influence and their inability to do so, comparing their feelings in this respect to those attending an attack of incubus or nightmare.

The ultimate effects of most anæsthetics are sedative or depressing in their character, for, while under the early influence of their exhibition, the action of the heart is increased, as shown by the more rapid pulse, later on, the pulse usually falls below its normal number of beats. With nitrous oxide this is not the case. Under its influence the pulse is more rapid, and remains more or less so during the continuance of the effect. Proof of this is afforded in the careful recording of the action of the pulse in one hundred cases by Dr. J. D. Thomas of this city. In one case it rose to one hundred and forty-four, in several others to one hundred and twenty-eight, while in a few there was little or no change.

A notable contrast between nitrous oxide and ether or chloroform is afforded in their after effects. The administration of ether is very generally and chloroform quite frequently followed by nausea, while both are about equally attended with great subsequent debility of the patient, frequently lasting for hours and sometimes for days. This is

PHYSIOLOGICAL ACTION. 13

in obedience to a well-recognized principle in the action of medicines in general.

Dr. George B. Wood, in his Pharmacology and Therapeutics, says: "One of the laws of all stimulation, whatever may be its degree, is that it is followed by a depression proportionate, at least proximately, to the previous exaltation of the function or functions excited."

Nitrous oxide, in its action, seems to set aside or evade this law, for its exhibition, when properly directed, is not followed by any noticeable depression. The patient, after his quick recovery from its influence, feels as well and strong as before, and is able to attend to his usual duties without any loss of time or feeling of languor. Nausea, too, following its administration is almost unheard of, even though given after a hearty meal. The author, in his extended experience with this agent, can recall but three cases where nausea followed the inhalation of the gas.

It has been charged against the gas by a few (the late Prof. George T. Barker among the number) that its after effects, although not serious, were often uncomfortable and long-continued, manifesting themselves in a form of cephalalgia and general debility.

It is believed, however, by those who have most carefully watched the effects of the gas through a series of years, that in the very few such cases brought to light the unpleasant results were attributable either to nervous shock following extensive operations on patients of delicate nervous organization; to the impurity of the gas; to its injudicious administration in too great quantity, or to the inhalation of carbonic acid gas exhaled from the lungs into a rubber bag from which the nitrous oxide was being given.

The two principal theories that have obtained in regard to the physiological action of nitrous oxide are *hyper-oxidation* and *hypo-oxidation*. The advocates of the former theory believe that the anæsthetic effects of the gas are due to the fact that in its administration the lungs are supplied with a greater relative volume of oxygen than they are accustomed to receive, and this being carried to the brain by the blood produces there a condition of over-stimulation, which exalts or raises the nervous centres above the point of sensation. They base their belief on the fact that nitrous oxide contains more oxygen than atmospheric air, and oxygen in its action produces a stimulating effect. This stimulating effect, which to them argues a superabundance of oxygen in the blood, is noticeable in the first stages of anæsthetization by this agent.

Those who hold to the theory of insufficient oxidation claim that the gas acts principally by occupying the space in the lungs normally held by the oxygen of the air, and to its exclusion, thus preventing the proper oxidation of the blood, followed, in consequence, by suspended sensation, muscular inaction, and the production of a condition in the system closely related to asphyxia.

The change in the color of the blood during the administration of the gas, as shown in the lips, eyelids and face, shows conclusively, they say, that the oxidation of the blood has been interrupted or interfered with.

The first theory had many supporters in the earlier days, but they rapidly fell away when it was shown to be incorrect. It was based upon the belief that nitrous oxide was a supporter of combustion at normal temperature, which proved to be a fallacy. It doubtless would be a supporter of combustion if it were a mere mixture readily separable; but, being a chemical compound, with properties peculiar to itself, and not resolvable into its original elements except at a high heat, it could not support combustion by virtue of the amount of oxygen it contained.

Dr. Thomas W. Evans, of Paris, writing upon this subject, says: * "The fact that nitrous oxide contains a greater proportion of oxygen than atmospheric air is no evidence, even *a priori*, that it possesses a greater oxygenating activity. The deutoxide of nitrogen is, as compared with nitrous oxide, doubly rich in oxygen; but it is not only immediately fatal to animal life, but is even incapable of supporting combustion. The physical properties of nitrous

---

*Dental Cosmos, Vol. XI., p. 286.

oxide differ widely from those of oxygen, as well as from those possessed by any mixture of oxygen and nitrogen, and the physiological effects of the gas are equally distinctive."

The theory of insufficient or hypo-oxidation is the one having the greater number of supporters at the present time. It has superseded the previous theory on account of its greater reasonableness and because the physical facts of the case support it.

If nitrous oxide is not an oxygenating agent or supporter of combustion, its presence in the lungs, to the exclusion of atmospheric air, would tend to prevent oxidation of the blood, and this would cause all the blood to circulate as venous, producing lividity of the countenance and gradually stupefying the brain and nerve centres so that narcosis or insensibility would supervene.

Now, these effects are all noticeable to a certain extent in the progress of anæsthetization, and it is but reasonable to presume that they are partly produced by the gas in the manner described.

Prevention of oxidation of the blood, however, would not be sufficient to account for all the manifestations attending the administration of the gas, and it has therefore come to be believed that the agent also produces a specific stimulating effect, upon certain nerve centres, which has not been clearly defined.

As showing the opinions held by recent authorities in regard to the physiological action of nitrous oxide, we give the following:

Prof. George Johnson says: * "Nitrous oxide is a rapidly acting anæsthetic, causing complete unconsciousness in less than a minute. At a high temperature it is an oxidizing agent, but at the temperature of the body it gives up no oxygen, but is exhaled again unchanged. When inhaled in place of atmospheric air it rapidly replaces the oxygen of the blood, and this being done, the functions of the brain are completely suspended and there is a state of profound coma, which quickly passes off when air is again allowed to enter the lungs." Dr. Robert Amory coincides with this view.

Prof. H. C. Wood, Jr.,† says that in anæsthetization by nitrous oxide there is "partial capillary stasis in the brain and diminished oxidation of the blood."

Dr. Turnbull ‡ says: "After numerous experiments and observations on man and animals, I have arrived at the following conclusions:

1. Nitrous oxide gas has a very limited range when given alone, owing to the rapidity of its action and still more rapid elimination.

2. It acts directly upon the cerebrum and muscular apparatus almost simultaneously.

3. It produces regular and progressive modification in the action of the heart and capillaries of the

---

*Medical Times and Gazette, April 3, 1869; Turnbull, p. 178.
†Dental Cosmos, Vol. XIII., p. 207.
‡ Anæsthetic Manual, 2d. Ed., p. 198.

skin, and, if carried to a greater extent, it affects the spinal axis, and lastly the cerebrum and medulla oblongata with suspension of respiration and circulation, and finally death.

4. Death in no case occurs without premonitory symptoms, and if respiration should cease for even a half to one minute, resuscitation is yet possible."

Dr. J. W. White, in an editorial in the Dental Cosmos * on the death of a lady alleged to have been caused by the inhalation of nitrous oxide gas, gives a very clear and concise statement of his views as to the physiological action of the gas. He says : " The conviction of the writer, based on personal experience repeated hundreds of times, as well as on observation and reflection, is that nitrous oxide, when inhaled, acts primarily by a specific stimulant effect on the centres of innervation (overstimulation and consequent depression if continued) and secondarily by preventing the oxygenation of the blood. That the inhalation of nitrous oxide, if continued, produces, by some method of action, no matter what its primary effect, progressive depression of vital functions, which tends to death, and in which the anæsthesia or temporary unconsciousness sought, is a more or less clearly defined step in the downward path there is no doubt. Immunity from danger can at the best be assured only by an intelligent and watchful guard, and its

---

*Vol. XIV., p. 311.

exhibition should be suspended while yet the centres governing respiration and circulation are not too profoundly impressed."

One feature connected with the anæsthetic state, of vital importance to the operator, is the effect that the various agents have upon the urino-genital organs. Fortunately, this effect is not often produced, or, at least, manifested, but when it is it may be the means of involving the operator in serious difficulty, unless he has taken the precaution to protect himself by the presence of a third party.

As in dreams during sleep, where portions of the brain are active while others are at rest, so in anæsthesia, where it is not profound, the patient in most cases has dreams of one sort or another, the character of which no one can foresee. In many cases they are erotic in character, and may so strongly impress the mind of the patient as to produce the conviction, upon restoration to consciousness, that the circumstances were real and not imaginary. In this way a perfectly honest patient may place an equally upright operator in a most embarrassing situation by accusing him of that of which he is entirely innocent. Cases of this kind have occurred, and are on record, resulting, in certain instances, in legal proceedings which proved detrimental to the character of the operator and materially injurious to his professional interests. These are alluded to more fully in chapter XII.

These hallucinations, of which the operator must

ever stand in great fear, have occurred in connection with the administration of all the principal anæsthetics, but whether they occur more frequently under the influence of nitrous oxide, we have no means of ascertaining.

Prof. Barker * says he never met with a case in which the aphrodisiac effect was noticeable.

Ziegler says: † "Nitrous oxide has a special tendency to these (urino-genital) organs and exerts a powerful influence over their functions." Again, he says: "Moreover, through its powerful aphrodisiac effects, it may intensify sexual desire to such a degree as to cause unpleasant exposure or even serious trouble."

Prof. Stellwagen, in his American revision of Coleman ‡ quotes Dr. Johnston, of Brooklyn, as follows: "Anæsthetics do stimulate the sexual functions. The ano-genital region is the last to give up its sensitiveness. Dentists or surgeons who do not protect themselves by having a third person present do not merit much sympathy."

The author, in the course of his personal experience, has met with a few cases in which the erotic effect was most marked. In one instance the patient (a man) by his actions gave unmistakable evidence of the vile character of his dream. Such

---

*Instructions in Nitrous Oxide.
†Researches on Nitrous Oxide, pp. 23-51.
‡Coleman's Dental Surgery, p. 336.

PHYSIOLOGICAL ACTION.

effects are, of course, due to the stimulative effects of the anæsthetic.

Another result, less serious in character, though quite undesirable and annoying, sometimes follows or attends the administration of an anæsthetic. It is the voidance of the contents of the bladder or rectum, due to the muscular relaxation of their sphincters.

Almost every anæsthetist has met with occasional instances of this kind, and they are more common with children than adults on account of the irregularity of their habits. As there is less muscular relaxation produced by nitrous oxide than by ether or chloroform, such accidents are more likely to be met with in the administration of the latter agents. As a preventive measure, in the case of children, it is well to request the parent or attendant to see that the organs are evacuated before the administration.

CHAPTER IV.

RELATIVE SAFETY.

In considering the safety of anæsthetic agents, and of nitrous oxide in particular, it must be borne in mind that the anæsthetic state is a pathological and not a physiological one. In entering it the patient gradually passes out of his natural condition into one that is unnatural and abnormal. The vital functions are altered and interfered with, and such interference, as in disease, constitutes a departure from the basis of safety. If such departure be slight or limited no ill result will usually follow, but if it be carried to a great length serious consequences may and often do result.

Dr. Squibb, one of our best authorities on anæsthetics, says: "The line of greatest safety in practice is to regard the difference between anæsthesia and death as a difference in degree only. The condition may be partial, full, profound or fatal, but with no distinct boundary lines between the degrees. The two intermediate degrees or stages constitute anæsthesia proper, and the full anæsthesia is generally required in surgery, while the stage of partial anæsthesia is generally sufficient in medicine."

RELATIVE SAFETY. 23

Bearing in mind the fact that all anæsthetics have associated with them the element of danger, and that when given it must be with a full understanding of the possible serious results, it follows that they should not be administered carelessly or by anyone who does not thoroughly understand their properties and action.

Of the various anæsthetic agents thus far introduced, nitrous oxide is on all hands conceded to be by far the safest. It has undoubtedly been used the greatest number of times; has probably saved the greatest aggregate amount of pain, and has produced by far the fewest deaths.

In 1870 Prof. E. Andrews * gave the following estimate of the relative danger from different anæsthetics:

Sulphuric ether, 1 death to 23,204 administrations.
Chloroform, 1 " " 2,723 "
Mixed chloroform
 and ether, 1 " " 5,588 "
Bichloride of
 Methylene, 1 " " 7,000 "
Nitrous oxide, No deaths in 75,000 "

One of the best evidences of the relative safety of nitrous oxide is the fact that in one office in the city of Philadelphia (Colton Dental Association succeeded by Drs. F. R. and J. D. Thomas) the gas has been administered for the extraction of teeth and minor surgical operations one hundred and

---
*Chicago Medical Examiner.

forty-seven thousand times without a single death or serious consequence.

During the past twenty-two years the author has administered it many thousands of times without a single unfavorable result. Its introduction and use have grown year by year, until there is scarcely a village in the more populated States of America where it is not administered daily in one or more dental offices, while in all of the larger cities there are offices wholly devoted to its administration. Considering then its wide-spread use, the varying ages and physical conditions of those to whom it is administered, and the fact that in many cases those using it have had no proper instruction as to its properties or manner of administration, we cannot but wonder at the few ill results that have followed its use. Up to the present time but four deaths have followed its administration in this country, and, as all of these occurred during the time that the gas was made in the office of the one using it, it is more than probable that some, if not all of them were attributable to impurities in the gas, resulting from imperfect preparation. So far as the author has been able to ascertain, no deaths have as yet resulted from the use of the liquified gas.

Dr. Squibb, in a lecture on anæsthetics before the Medical Society of the State of New York, says: "Nitrous oxide was the first anæsthetic, and the safety and certainty of its effects, even in inexperienced hands, for all momentary operations, and the promptness with which persons recover from its effects, render it, perhaps, the most important of all anæsthetics, because destined to relieve a greater aggregate amount of pain, with greater safety, than any other agent."

Dr. Turnbull * says : "Nitrous oxide is the safest of all anæsthetics."

Mr. Underwood, of London, says : † " Nitrous oxide gas is the best anæsthetic at present known to the profession. The risk to life is so small that it may safely be said that, supposing a cardiac condition existed that rendered nitrous oxide gas a dangerous agent, in such a case any operation, even the extraction of a tooth, *without an anæsthetic*, would be attended with still greater danger. To put the case in other words, every short operation becomes less dangerous to life when performed under gas than when the anæsthetic is not employed. The gas, administered with ordinary care by some one whose entire and undivided attention is devoted to its administration, renders less, and not *greater*, the risk to life, if any such risk be supposed to attend the extraction of a tooth."

---

*Op. Cit., p. 204.
†Notes on Anæsthetics, 1st Ed., pp. 12-13.

CHAPTER V.

ADVISABILITY OF ADMINISTRATION IN SPECIAL CASES.

Before entering upon the subject of the administration of nitrous oxide gas, it may be well to consider first certain abnormal or other conditions of the system in which the advisability of its exhibition might be questioned.

Foremost among these are those relating to the heart and lungs. Any affection of these important organs, while it might not contra-indicate the use of the agent, would certainly introduce an element of possible danger that should be thoroughly understood and, as far as possible, guarded against. The administration of the gas to patients affected with valvular disease of the heart or phthisis has always been considered inadvisable, although the rarity of a death directly traceable to either of these conditions, and the further fact that the gas has undoubtedly been administered in many cases where these conditions were not recognized, have caused them to be considered less dangerous than they once were.

Underwood says : * " Heart disease is no draw-

---

*Loc. Cit., p. 29.

back to the administration of nitrous oxide. Very fatty, weak hearts have been supposed to expose the patients to some risk, but there has never been a fatal case traceable to this cause. Moreover, as this condition is not diagnoseable during life, it is futile to consider it. Organic disease does not involve any additional risk whatever to the patient."

Coleman says: † "Patients who have suffered from acute rheumatism, resulting in damaged valves of the heart, etc., appear to take nitrous oxide as well as ordinary patients, but the appearance of a lady affected with cyanosis was such that we should in future decline to administer it in similar cases. Those with weak and fatty hearts must ever be unsafe subjects for the gas or other anæsthetics. Should we have a knowledge or suspicion that we have to deal with a patient so circumstanced, we should, besides redoubling our precautions with regard to careful observation of all symptoms, pay especial attention to the conditions of the pulse. We strongly recommend operating only whilst the patient is fully anæsthetized," for the reason that there will be less chance of arresting the heart's action through shock."

Turnbull says: * "Those who are less skilful and inexperienced should reject cases of great physical exhaustion or patients with a feeble or fatty heart."

---
†P. 336.
*P. 167.

While it is most important to ascertain beforehand, either through our own examination or that of a physician, whether there be any abnormal condition of the patient's heart, we need not be deterred from the administration by the patient's own statement of the condition of his or her heart, for the average patient, when questioned, will in nearly all cases confess to some derangement of that organ. Palpitation, caused by excitement or slight exertion, is most common and harmless, and yet the uninformed patient will be most likely to construe the condition into a serious one peculiar to himself. Such persons can generally be best comforted and assured by placing one's ear or the stethoscope to the patient's chest and noticing the character of the pulsation.

In *phthisis* the danger will be more or less great according to the extent of the disease. In the earlier or middle stages there will be no real danger, but the patient, owing to debility and the obliteration of lung tissue will require less than the ordinary quantity of gas to produce anæsthesia. In advanced stages of the disease it is better to avoid the administration, if possible, but if the suffering of the patient can be alleviated in no other way, we should give it with the greatest care, noticing well the respiration and pulse, carrying the patient only to the stage of insensibility and not beyond it. That it is proper to give it, as a last resort, under the last mentioned circumstances, cannot be questioned, for the shock attending extraction without the gas would be more likely to prove fatal than its administration.

Coleman says: * " Phthisical patients take nitrous oxide fairly well, becoming anæsthetized with less gas than ordinary patients, as we should have anticipated. Anæmic patients do the same."

In patients of *apoplectic* tendency or appearance, those of full habit, with short, fat necks, there will usually be no difficulty, provided the clothing about the throat be loose and the head be kept as nearly as possible on a line with the body. Respiration is more easily interfered with in these than in those of slender form.

Epileptic persons can inhale the gas without danger. A seizure may precede or follow the administration, but, of course, cannot occur during it. In either case we need but follow the usual course of placing the patient in a recumbent position, loosen the clothing, and see that there is not too long a suspension of heart action. On one occasion in the author's practice, a girl was seized with an epileptic fit as he was about to apply the forceps for the extraction of a tooth. After recovery the gas was administered and the tooth extracted with the most satisfactory results.

Hysterical patients, while recovering from the influence of the gas, often give way to their pre-

---

*P. 336.

vailing emotions in a manner calculated to alarm a novice, but experience will show that no danger need be apprehended from their manifestations.

Alcoholism is a condition that the author has found to be very unfavorable for the administration of the gas. If the patient be a confirmed drinker, no untoward results may follow, but a greater quantity of the gas will need to be given in order to produce anæsthesia. If he be only an occasional drinker, and happens to be somewhat under the influence of liquor at the time, gas should be denied him, as most likely it will only serve to increase his already excited condition, and he may do violence to the operator or to himself. One experience of this kind will be sufficient for a lifetime, as the author knows from a case in his own practice.

Idiosyncrasy.—This peculiarity is so common, and its manifestations so various, that it would be difficult to treat it in detail. The administrator of gas often encounters it in persons who declare that they never have been and cannot be brought under its influence. That such a condition could exist we may not be justified in denying, but with pure gas, properly administered, we do not see how any ordinary constitution can successfully resist its influence, though some, as we know, are more susceptible to it than others. Many operators of largest experience in the use of gas declare that they have never met with a case where anæsthetization was impossible, although in some cases those anæsthetized declared in advance that they could not be brought under the influence.

Menstruation has proven to be no bar to the administration of the gas. No ill results have been known to follow on such occasions.

Pregnancy, not later than the seventh month, has been found not to be interfered with by the extraction of a few teeth under the influence of the gas. Later than the seventh month might be safe also, but the author has never attempted it and has always advised against it. Extraction, with or without an anæsthetic, should be postponed, if possible, until after confinement.

Lactation does not contra-indicate the exhibition of the gas. Mr. Braine* mentions two cases in which the shock attending extraction of a tooth without the gas stopped the secretion of milk, whereas the same patients underwent similar operations a little later, under the influence of the gas, without any ill results.

Age seems to present no barrier to the use of nitrous oxide. It may be given (other conditions being favorable) as early as the child can be made to inhale properly or quite late in life. Mr. Braine states that he has given it to a patient at the age of ninety-four. The author has administered it to patients three years of age and again to those of nearly eighty.

---

*Journal of British Dental Association.

CHAPTER VI.

MANUFACTURE.

The apparatus required for the manufacture of nitrous oxide consists of a retort or alembic, several wash bottles and a receiver or holder.

The retort should be of strong glass, fitted with a ground-glass stopper at neck, where the charge is introduced. In use it should be seated in a sand-bath, which, in turn, should rest upon a gas stove. To maintain its upright position the retort should be supported at its neck by attachment to an arm of a retort stand, or other suitable appliance.

Fig. 1.

The wash bottles should be of the form known as Wolfe's bottles (having two or three necks,) or halfgallon bottles with a single wide mouth fitted with a rubber cork having two perforations for the reception of the bent glass tubes as shown in Fig. 1. In two of the bottles, the longer tube should reach nearly to the bottom of the bottle, and should preferably terminate in a bulb pierced with numerous holes for the breaking up of the column of gas into bubbles to facilitate the process of washing or purification.

The holder or tank, together with the receiver, may be made of sheet zinc, galvanized iron or copper (nickel-plated). The receiver, which is open below, should fit loosely into the holder and be supported by means of cords and pulleys attached to an upright framework which is fastened to the holder. The free or outside ends of these cords should have attached to them suitable weights to counter-balance the weight of the receiver. The holder should have a capacity of about fifty gallons. A metallic tube of one inch diameter should extend from the upper edge of the holder along its outside to the lower edge, thence along the bottom to the centre, there entering the holder and extending upward to a level with the top. The upper outside termination of this tube is supplied with a faucet which is closed when the gas is being manufactured, and opened when it is to be inhaled. This same tube has another faucet at the base of the holder which is open during the elimination of the gas and closed during its administration. Another faucet is attached to the holder at some point of its base for the purpose of emptying it when this becomes necessary.

Wash bottle No. 1 (see cut,) should be empty, as it is simply a drip-bottle, and the tubes need not ex-

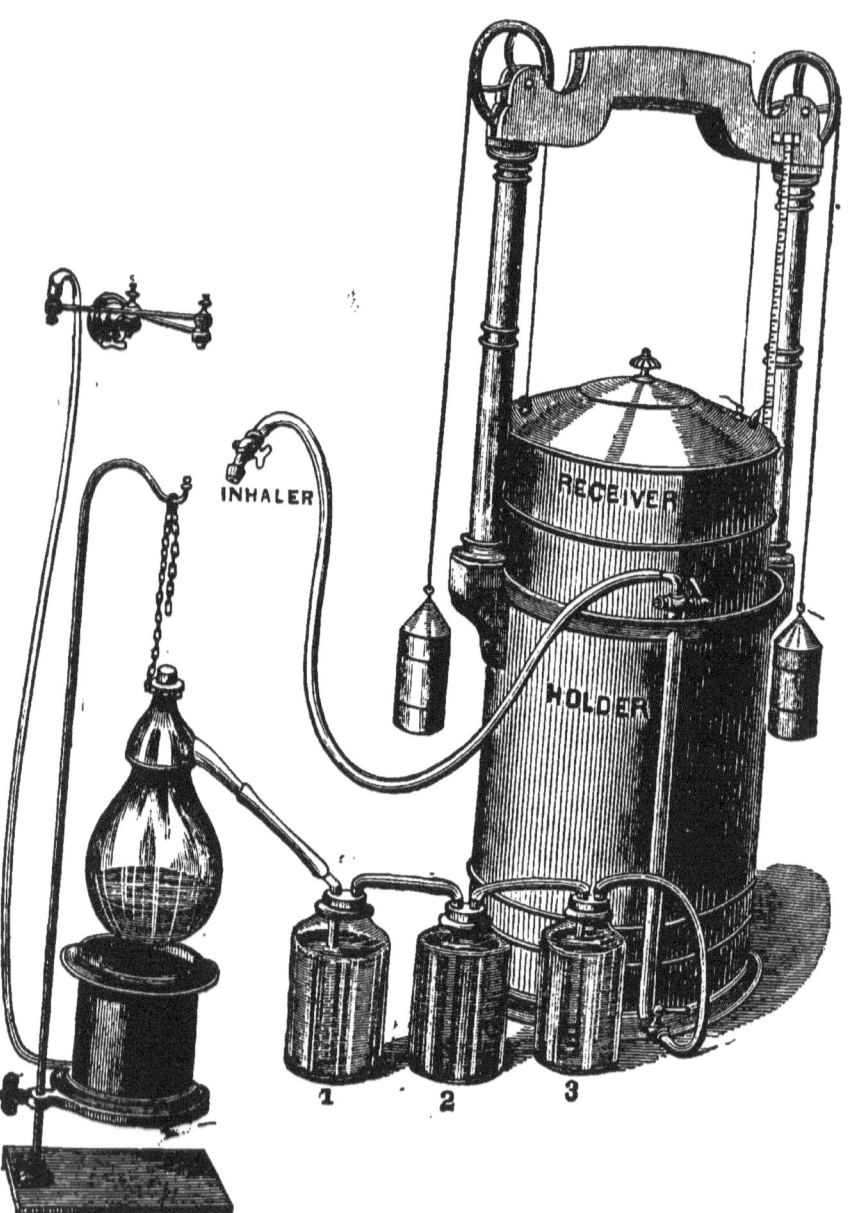

Fig. 2.

tend far through the cork. Bottle No. 2 should contain a quart of water in which four ounces of ferrous sulphate have been dissolved. Bottle No. 3 is charged with a solution of one and a half ounces of caustic potassa in a quart of water. The bottles should be connected with each other (the short tube of one with the long tube of the other), and to the retort by means of rubber tubing. A rubber tube should also connect the short tube of bottle No. 3, with the lower tube faucet.

The holder should be filled to within a few inches of the top with pure soft water. Removing the weights from the cords and opening the upper faucet of the holder, the receiver will settle to the bottom of the holder by its own weight, after which all faucets should be closed and the weights readjusted.

The retort, charged with about one-and-a-half pounds of ammonium nitrate, (the granulated being the form most convenient for introduction) should be placed in a sand-bath resting on the gas stove, and after being connected with wash bottle No. 1, heat may be applied.

As soon as the salt begins to melt, the lower tube-faucet should be opened to permit the entrance of the gas into the receiver. The elimination of the gas will begin as soon as the fused salt begins to boil or bubble, and its passage to the receiver will be indicated by the bubbling in the second and third wash bottles.

The salt fuses at about 220° F., and at 470° to 480° the gas is evolved. When this point is reached the heat should be kept as uniform as possible until the close of the operation, to avoid the formation of the dangerous gases $N_2 O_2$ and $N_2 O_3$. The formation of the former is indicated by cloudy fumes in the wash-bottles, and the latter by the presence of orange-colored vapor. These latter gases are given off at a temperature of 500° and over.

When the desired quantity of nitrous oxide has been generated, the heat should be cut off, the tubing between the first and second wash-bottle detached, and the lower tube-faucet closed. As water absorbs about its own bulk of the gas, the first receiver-full will disappear if the weights be lessened. A second quantity will therefore have to be made before it can be drawn upon for administration. This absorption of the gas by the water, occurs of course only once after each re-filling of the holder with water. Such changing of the water need only take place at intervals of a month or two. The solution in the wash-bottles will remain active for weeks, and the necessity for renewal will be indicated by the formation of a red precipitate in the iron solution, and crystals of potassium nitrate in the potash solution.

The water in the holder, even though saturated with nitrous oxide, exerts a purifying influence upon the gas in the receiver. For this reason, gas should not be used immediately after it has been made, but should be allowed to stand for six hours or more before being administered.

It has been claimed by some that gas should only be used while "fresh," and for this reason should be manufactured daily. It has never been proven that gas undergoes any deterioration when contained in an air tight receiver over water, nor do we see how this can take place.

LIQUIFIED NITROUS OXIDE.

The time and care required in the manufacture of nitrous oxide, together with the bulkiness of the necessary apparatus, has led to its general abandonment since the manufacture and liquifaction have been entered into on a commercial scale by certain dental firms.

The manufacture of gas for liquifaction is conducted in the same manner as that already described, except on a larger scale. Instead of glass, iron retorts lined with porcelain are used, and every care is exercised both in the matter of preparation and purification to produce as pure an article as possible. For condensing the gas, a pump worked by steam is used and by it the gas from the gasometer is forced by slow and regular action into the iron cylinders and thus liquified. As previously

---

For eight years previous to the introduction of liquified gas the author manufactured his own gas and administered it daily. Occasionally, during a temporary absence, gas would remain in the receiver for two or three weeks, and when administered invariably produced as satisfactory results as gas only a day old.

stated a temperature of 45° F. must be maintained to bring about the result, and to this end during the charging of the cylinders they are kept in ice and water, which serves additionally to carry off the heat developed in the condensation.

After the cylinder is charged with its proper amount, the tight-fitting valve with which it is pro-

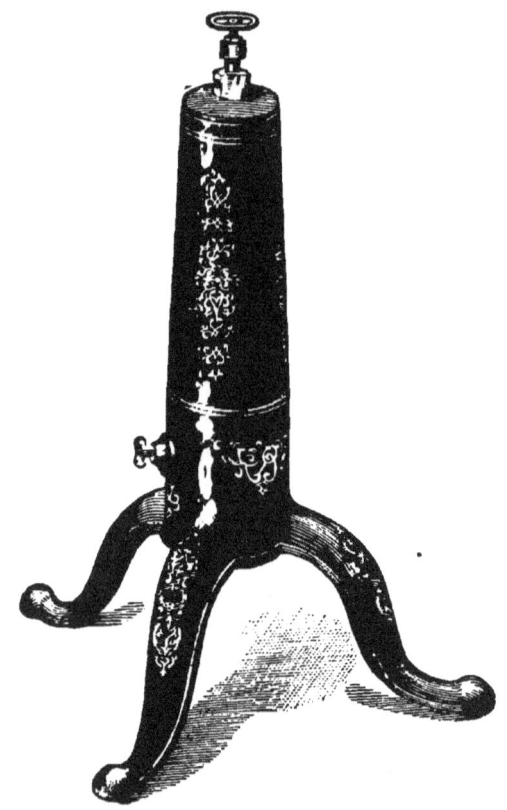

Fig. 3.

MANUFACTURE. 39

vided is closed, and the cylinder disconnected from the pump. About fifteen minutes are required to charge a hundred gallon cylinder and a proportionally longer time for one of larger dimensions. The cylinders are heavy wrought iron tubes with top and bottom welded in, and to insure tightness are tested under heavy hydraulic pressure before being used.

Gas in this form can be transported anywhere, and accommodation for the small cylinder is easily

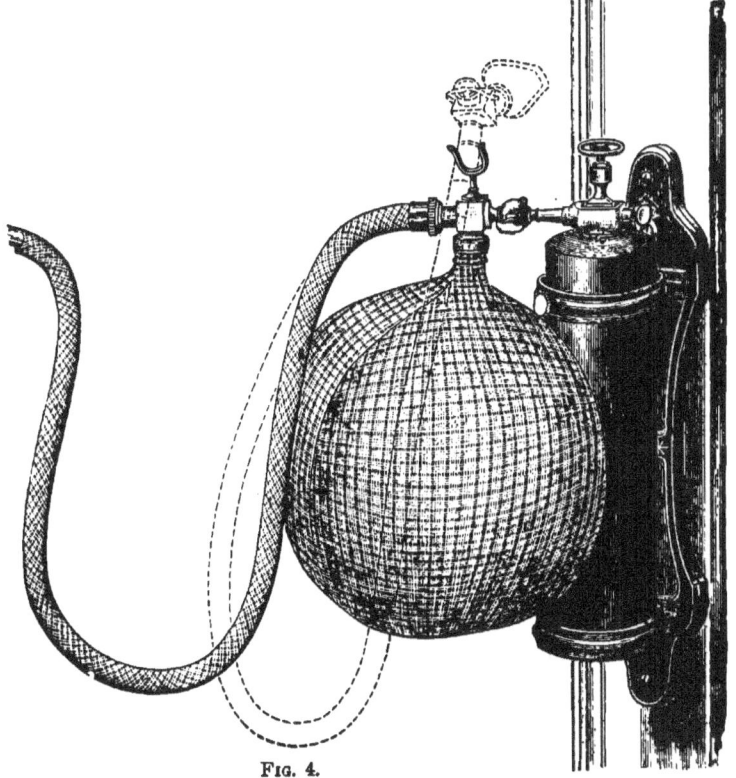

Fig. 4.

found in any office. In use it may be supported in a tripod or ornamental stand, (Fig. 3,) or attached to the wall by a bracket. (Fig. 4.)

For convenience in carrying to the house of a patient, or to a distance, a metal case, covered with morocco, and made large enough to contain not only the cylinder, but also the bag, inhaler and tubing, is preferred. (Fig. 5.)

Fig. 5.

Fig. 6.

In offices where the gas is administered on a large scale, it is often preferred to allow the gas to escape from the cylinder into an ornamental gasometer holding about ten gallons, and having the patient inhale the gas from the gasometer. An apparatus of this kind, with the cylinder attached, is shown in Fig. 6.*

It is claimed for the gasometer, that in addition to the convenience of the large quantity it may contain, the pressure of the receiver upon the contained gas is more uniform than where a bag is used, and the inhalation is rendered easier on the part of the patient.

Where gas is administered from an inflated rubber bag, as shown in Figs. 4, 5, the pressure during the earlier inhalations is excessive, whereas when the bag is nearly empty, quite an effort is required on the part of the patient to extract the last portions. This inequality of pressure is a decided disadvantage, and is entirely obviated by the perfectly uniform pressure from first to last of the receiver of the small gasometers. Another advantage derived from the use of the gasometer is economy, for the portion of gas unused after one administration is retained for future use. In inhaling from a bag, the residual portion, if there be any, is necessarily lost.

---

*All the foregoing cuts are introduced by courtesy of the S. S. White Dental Mf'g. Co., and represent articles of their manufacture.

MANUFACTURE. 43

Fig. 7 represents a gasometer devised by Dr. A. M. Long, of Monroe, Michigan, which is highly endorsed by those who have used it, and embodies many valuable features. Like the one represented

FIG. 7.

in Fig. 6, it consists of a water tank or holder in which is suspended an inverted receiver with a guide-rod in its centre to preserve its upright position. Both holder and receiver are made of thin seamless copper tubing, nickel-plated, which affords sufficient strength together with lightness. It differs from all others of its class in being provided with a metal float with soft rubber flange fitting the inner periphery of the receiver, thus seeming to combine the advantages of a dry gas chamber with a wet cistern and sensitive motion.

CHAPTER VII.

INHALERS AND ACCESSORY APPLIANCES.

One of the most important adjuncts for the administration of gas is a properly constructed inhaler. Even with pure gas and a perfect receptacle for it, its use may be fruitless of good results or attended with serious difficulties if the inhaler be faulty in principle or construction. The requisites of a good inhaler are, that it be perfectly air-tight in its fittings; that it be of sufficient calibre to admit a full volume of gas; that its valves be free in action and tight when closed; and that it be so formed as to fit the face perfectly. It should also be provided with a simple yet accurate cut-off to prevent the escape of gas at the close of the administration.

Any admixture of air occurring through imperfection of the appliance or carelessness of the operator will either very much delay or wholly prevent the desired result. The calibre of the tubing and inhaler is also of great moment. It should be large enough at every point to permit the passage of a volume of gas sufficient for full and easy inhalation on the part of the patient. The lividity of the countenance, and especially of the lips and eyelids, so noticeable in the earlier days of gas administration, was undoubtedly due to the fact that the patient received the gas at each inhalation in such limited quantity that partial asphyxiation took place before the anæsthetic stage was reached. Since the introduction of improved inhalers of larger calibre, this appearance of asphyxiation is almost entirely done away with. With the larger volume of gas available and the unlabored action of respiration the full anæsthetic stage is reached before asphyxiation can take place.

The first inhalers made were provided with a hard rubber mouth-piece. To prevent the inhalation of air in its use the lips were held in contact with it by means of several fingers of both hands of the operator, while the nostrils were closed by the pressure of some of the remaining fingers. This method was not only exceedingly inconvenient for the operator, but also uncomfortable as well as alarming to the patient by the feeling of suffocation produced. The rubber hood now most generally used is an improvement in every way, for while it adapts itself readily to the face, covering both mouth and nose and allowing inhalation through each, it leaves the hands of the operator more free for other service.

The accuracy and perfect working of the exhaling and inhaling valves of every inhaler is a matter of the greatest importance, for if they fail in the proper performance of their functions, either air will be admitted during inhalation, or exhaled nitrous oxide combined with carbonic acid gas will be forced back into the receiver to be again inspired with the next

Fig. 8.

inhalation. Fig. 8 represents the Codman & Shurtleff inhaler, which in the hands of the author has proven a very effective instrument. Letter B indicates the rubber hood, C the outlet and position of exhaling valve, E the position of inhaling valve, and D the two-way stop-cock with thumb lever.

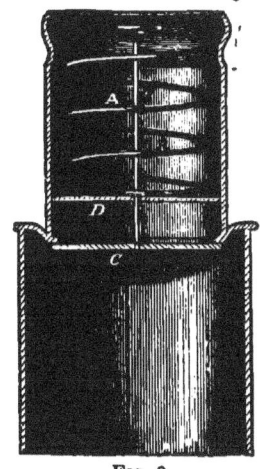

Fig. 9.

Fig. 9 illustrates the position and construction of the valves in this inhaler. The thin, hard rubber disk, C, rests against an annular metallic flange and is held in position and rendered sensitive by the delicate, hair-like spiral spring, A B.

A more recent inhaler, and one combining all the good qualities that our present knowledge could demand, is

48   NITROUS OXIDE.

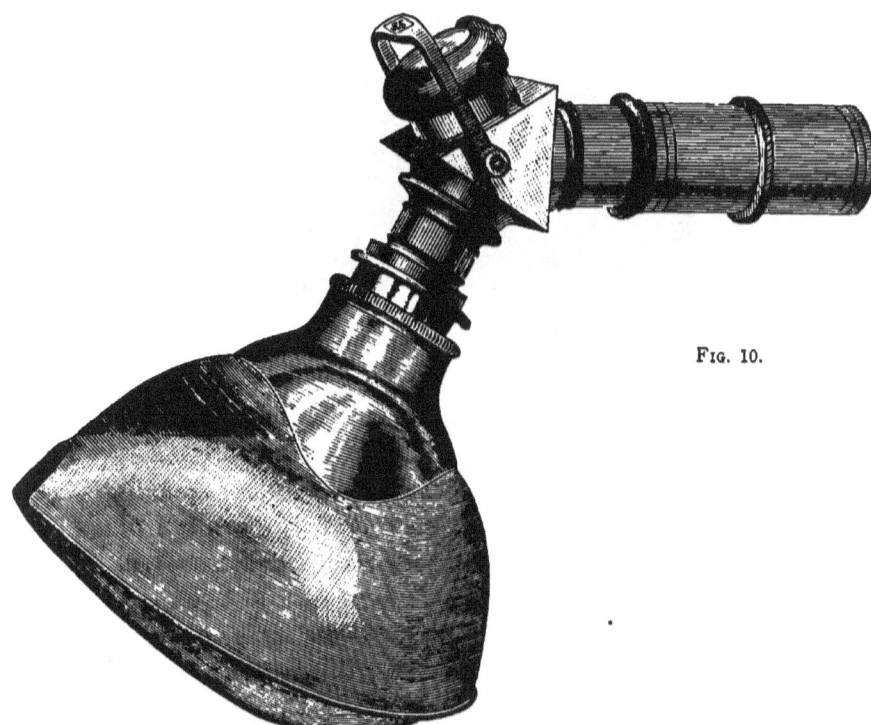

Fig. 10.

the one represented in Fig. 10, devised and manufactured by the S. S. White Dental Manufacturing Company. In it the character of the valves is very similar to the one previously described, but they are closer together, and the swinging lever on the outside operates a cam on the inside of the triangular box which effectually controls the action of both the inhaling and exhaling valves. It is provided with the rubber hood, the tubing is of large diameter, and the workmanship throughout is of the best character. Rubber hoods, with an inflatable margin to more perfectly fit the face, have been in use for many years in England (Mr. Clover having devised the first one) and more recently have been introduced into this country. They seem to possess an advantage over the plain hood in the greater range of adaptability.

Gags or mouth props are indispensable adjuncts in the administration of nitrous oxide gas, for with this agent more than with any other, is it necessary to keep the jaws mechanically distended during extraction. Gags, to be properly efficient and safe, should possess the following properties:

1. They should be strong and indestructible.

2. They should be easy to cleanse, hence non-absorbent.

3. They should keep their place and not slip.

4. They should be sufficiently elastic to prevent injury to frail teeth.

5. They should be large enough to be incapable of being swallowed, or be protected with a strong twine.

6. There should be a sufficient variety of sizes to meet all cases.

Gags have been made from large bottle corks cut to shape, but they are dangerous to use, owing to their ready destructibility and the possibility of portions of them finding lodgment in the trachea. In addition to this, being absorbent, they are necessarily unclean. This last objection holds equally good in reference to those made of wood or felt. Wooden gags are also objectionable because they are liable to injure frail teeth or those containing large fillings in the muscular pressure brought to bear upon them. Metal or hard rubber gags, faced on the ends with elastic rubber, have found favor with many, but all those that the author has met with were so slender in their short diameter as to seem likely to become dislodged by any lateral movement of the jaws.

Many years ago the author devised and constructed a set of three, which he has had in continuous use ever since, and which so fully meet all the requirements of a gag that he has met with nothing better. Fig. 11 represents the medium-sized one. They are made of partially vulcanized rubber, such as is used for car springs and carriage buffers. The material can be bought in bulk at the rubber stores and cut to shape and made smooth with a knife and vulcanite file. As will be seen by the cut, the two transverse diameters are nearly equal, while the long diameter varies in each one. Two sides, those presenting to the tongue and cheek, are flat, while each of the other four sides is concave. These concave surfaces are to accommodate the teeth or the alveolar ridges, and by their shape lateral slipping is entirely pre-

Fig. 11.

vented. Each gag, by being made concave on the ends and two sides, has two heights, either one of which may be used. By virtue of this the three gags have a range of distension varying from three-fourths to one and three-eighths inches. They can be readily and perfectly cleansed, are indestructible, and, while the teeth may partially compress them, they are incapable of injuring the frailest tooth. Their size precludes all possibility of being swallowed, and hence there is no need of a string to prevent any accident of this character. When they are to be removed a bent finger will easily accomplish it. Their size also admits of at least two teeth in each jaw resting on their surface, which adds materially to their fixity of position.

An apron to protect the clothing of the patient is necessary and is preferably made from either white or black rubber cloth. It should be two feet wide by three and a half feet long, cut out at one end to fit the neck and provided with tapes to tie at the back.

A hand spittoon is useful for the relief of the patient before he is able to use the one by the side of the chair. A very neat and suitable one is made by the S. S. White Co., a cut of which is shown in Fig. 12. It should be placed near at hand before the gas is administered.

Fig. 12.

CHAPTER VIII.

ADMINISTRATION.

Preceding the administration, certain precautions should be taken to insure the best results. Artificial teeth, if worn, should be removed. Ladies should be instructed to loosen their dress or stays if they be at all tight, so as to give free play to the organs of respiration. If there be any friend of the patient present, she should not be allowed to remain by the chair during the operation, for, should she become alarmed at the appearance of the patient, she may, by incautious words or actions, excite the patient or annoy the operator. If an assistant be present, the friend should remain in an adjoining room. If the operator be alone, the friend may be allowed to remain in the operating room, as far as possible removed from the scene of operation.

It is of the utmost importance that absolute silence be observed by everyone present during the administration. A few words of instruction or encouragement addressed to the patient by the administrator, in a gentle tone, during the first stages of anæsthetization are sometimes beneficial, but beyond this no conversation should be held. Any communication between those present, when necessary, should be made by gesture or facial expression.

Some operators are accustomed to speak to their patients during the whole process of anæsthetization, thus trying to direct their thoughts into pleasant channels in order that their dreams may be agreeable, but the majority of those experienced in the administration of anæsthetics, have found it best to allow the thoughts or impressions of the patient to follow their own course. Nothing is more soothing at such times than silence.

In all cases it is important that an assistant be present to aid the operator whenever necessary. Some writers insist upon having the assistant administer the gas, so as to leave the operator entirely free. We see no necessity for this; indeed, we prefer to be both the administrator and operator, employing the assistant only to watch the pulse and respiration, to take charge of instruments or accessories as they are discarded, or to control the movements of the patient, should this become necessary.

By way of preparation for any emergency that may arise, the operator should have close at hand and within easy reach certain accessories. An artery or tongue forceps to assist in drawing out the tongue in cases of threatened asphyxiation, is a valuable adjunct. A small napkin, with which the tongue may be easily grasped, drawn out and securely held, will often prove serviceable where there is room to insert the fingers. Nitrite of amyl is invaluable in hastening the return to consciousness and quickening the action of the heart. Aromatic spirits of ammonia is also useful to apply to the nostrils where recovery is slow.

All of these, together with any instruments that may be needed during the operation, should be kept out of sight of the patient to prevent any unnecessary alarm. Forceps especially should be kept carefully concealed until the patient has lost consciousness. The less display of appliances the better for both patient and operator.

When the gas is to be administered the patient should be made to occupy a position in the chair comfortable to himself and favorable to the operator. A careful examination of the mouth should next be made and a mental note taken of the different teeth to be extracted, their position and surroundings, and their relative ability to stand the pressure of the forceps without breaking. The operator should also decide the order in which each should be removed. This being done, while the assistant is arranging the apron and other details the operator should open the instrument drawer and quietly arrange his forceps in the order in which they will be needed. At the same time he should consider the possibility of breakage or other accident, and, to meet such emergencies, should also arrange in proper order any accessory instruments that may be needed, after which the drawer should again be closed. With proper care all this may be done without attracting the patient's attention.

The gas should now be drawn from the cylinder into the intermediate bag or gasometer. Before beginning the administration, the operator should endeavor to calm any fears the patient may have by assuring him of the safety of the agent and explaining to him the pleasant and delightful sensations he will probably experience during the inhalation. He should also be instructed to breathe slowly and naturally and to make deep and full (though not labored) expiration and inspiration. If timidity be still noticeable in the patient it is often well to apply the inhaler to one's own face, and, by inhaling and exhaling through the exhaling valve, illustrate the manner of the operation.

The apron having been adjusted, and all being in readiness, the prop or gag should be introduced and placed in such position as to keep the mouth well open and be as much as possible out of the way of the operator.

The face-piece or hood of the inhaler should now be adapted to the face of the patient and he be requested to breathe. For a few seconds, or until he has acquired the proper manner of respiration, the patient should be allowed to breathe only the air through the exhaling valve. This will inspire confidence and allow time for the further allaying of any fear. This accomplished, the lever controlling the inhaling valve should be gently and cautiously opened and the gas admitted. If all goes well, as it should, the patient will continue the breathing without interruption until the anæsthetic stage is reached. It sometimes happens that the patient, when beginning to take the gas, will make labored movements of the chest simulating those of respiration, will grow red in the face, and then, forcibly removing the inhaler, complain of a feeling of suffocation and an inability to breathe. If the valves of the inhaler be all right and there be no interference with the flow of gas, we may be assured that the difficulty complained of rests with the individual and not with the apparatus. It is caused by the involuntary closure of the larynx, brought on by fear or lack of confidence; this should be explained to the patient, and an effort made to restore confidence, after which, upon a second trial, all will probably proceed satisfactorily.

Should the patient at any point of the procedure offer resistance to the continuance of the administration, he should be gently cautioned to keep quiet and assured that all is well. Should this not avail, and consciousness be still present, it is better to cease the administration and allow the patient to recover. If force be used or an effort made to restrain the movements of the individual under the above conditions it will only serve to terrify him and increase his struggles for release. If, however, such movements of the patient should occur after consciousness is gone and when he is almost anæsthetized, his movements should be restrained by the assistant and the administration pushed to completion. Good judgment on the part of the operator is needed to determine the exact condition of the patient when struggling or resistance begins, so that it may be treated in the proper manner. In the one instance the resistance is caused by fear, in the other it is due to some dream or unconscious mental impression.

When the gas is exhibited under favorable conditions the character of the respirations at the outset will continue the same until near the close of the operation. As complete anæsthesia is approached the breathing will usually become more rapid, and this will be followed in most cases by spasmodic or jerky inhalations very similar to snoring in natural sleep. The countenance will also usually give indication of the progress toward the anæsthetic state. The natural color at the beginning will gradually give way to pallor, which will continue while the effects last. This pallor is also sometimes accompanied by a greater or less lividity of the eye-lids and lips, though this is less likely to be manifested when the supply tubing is of large diameter and the valves of the inhaler of such a character as to admit a full and free supply of gas. This lividity, as explained in a previous chapter, is due to lack of oxygenation of the blood and is the precursor or beginning of asphyxia. It usually does not occur until the snoring is noticed, and both of these, when present, are indications of complete anæsthesia and require the suspension of the administration.

Muscular twitching of the extremities is also frequently noticeable as the anæsthetic state is approached, but no special significance is attached to it.

Complete anæsthesia, when attained, may be recognized by certain diagnostic indications:

1. If lividity of the lips and eye-lids be noticed we may feel sure that unconsciousness is near at hand, and to carry the administration much farther would be hazardous.

2. When the snoring * becomes marked it affords us an infallible sign that the full anæsthetic stage is reached.

3. The passing of the hand or finger rapidly before the open eye of the patient, without touching it, will indicate to us by the non-closure of the eye-lid that unconsciousness is complete.

The first two indications are the most reliable and readily discernible, but if they leave us in doubt,

---

*The snoring occurring as the anæsthetic stage is reached should not be confounded with the stertor preceding death. The former is physiological, the latter pathological. Snoring arises simply from a relaxed state of the muscles of the palate region. Stertor takes place in the larynx and indicates that the recurrent pharyngeal branch of the pneumogastric nerve is affected.

the third will assist us in deciding when the administration should be suspended.

Possible Complications.—The two most serious complications that may attend or follow the administration of an anæsthetic, are the arrestation of the action of the heart or of the organs of respiration. In all cases these need to be watched for and carefully guarded against. To this end it should be the duty of the assistant to carefully watch the action of both organs and give notice to the operator whenever any unusual symptoms occur. In the early stages of the administration the pulse will rise considerably from the stimulative effect of the agent. Afterward it will gradually fall to or nearly to the normal standard. Should it fall below normality there will be just cause for alarm, and if, in connection with this, the breathing should become labored and diminished in frequency, a crisis is at hand which will require all our efforts to be devoted to the resuscitation of the patient. If taken in time, the application of ammonia to the nostrils or the inhalation of nitrite of amyl will speedily restore the patient to consciousness. Failure in respiration, when it occurs, will in nearly all cases be accompanied by increased lividity of the countenance, indicating asphyxiation. This is most generally caused by the dropping of the tongue into the fauces, thereby cutting off all inhalation. The proper remedy in such cases is to seize the tongue with napkin, tenaculum or artery forceps, and draw it forward out of the mouth. In all cases of interference with heart action or respiration the only safety lies in the prompt application of suitable remedies, and to this end all such means should be readily available. A moment's delay may prove fatal to the patient.

After-Effects.—Fortunately, the administration of nitrous oxide is seldom followed by any after-effects, and such as are sometimes encountered are not of a serious nature. Nausea will occasionally occur after recovery, but when it does it is usually due to extreme sensitiveness of the stomach or to the accidental swallowing of blood. In either case, when the first symptoms appear, we may lessen the violence of the attack, or prevent it altogether, by the administration of a little brandy and water, and by affording the patient plenty of fresh air.

In ether and chloroform administration, attention is generally paid to having the patient's stomach neither over-full nor empty, as a preventive of nausea, but with gas this precaution has not been found to be necessary, for experience has shown that it may be administered at any time of day and under all ordinary conditions of the stomach without greater liability to nausea in the one case than in the other.

Syncope has sometimes, though rarely, been found to follow the administration of gas, but it is due in nearly all cases to a predisposition to the ailment or to the excessive weakness of the patient from some constitutional cause.

CHAPTER IX.

EXTRACTION DURING ANÆSTHESIA.

Extraction during the anæsthetic state is an operation requiring special qualifications on the part of the operator. He should be skillful in order that he may perform the operation in the best manner, with the least injury to the surrounding parts and least liability to accident. It has been well said that no one should undertake extraction under an anæsthetic until he has acquired considerable skill in extraction without it. He should be possessed of strong and steady nerves, so that he may not lose control of himself through excitement. He should be quick in his movements to enable him to perform the task he has set himself during the continuance of full anæsthesia. A clear head, steady nerves, and dexterity are all prerequisites to success in this branch of practice.

Unlike operations in general surgery, where anæsthesia is maintained during the continuance of the operation, the dental practitioner is obliged (from the location of the organs operated upon) to cease the administration before his work can begin, so that the anæsthetic state is rapidly passing away while his operations are being carried forward. For this reason no time must be lost or false movement made.

The position of the operator should be to the right of the patient, as it affords him greater facility and enables him to firmly grasp the patient's head with his left arm and thus steady it. This position should never be more than slightly altered, and the forceps should, if possible, be of such shape as to enable him to grasp any tooth or root in the mouth from his one position. Standing in front of the patient to grasp a molar on the right side of the lower jaw, as is sometimes done, places the operator at a disadvantage in the application of his force and prevents him from steadying the head with his left arm. During anæsthesia the patient's head should never be allowed to pass from the control of the operator.

While the position of the patient in the chair is practically the same under all circumstances, the relatively high or low position of the chair has much to do with facilitating the operation. Where lower teeth only are to be extracted the chair should be low in position to enable the operator to apply his power to advantage. This he can best do by standing over the patient and slightly back of him.

Where upper teeth only are to be extracted the chair should be so arranged as to bring the patient's head about on a level with the operator's breast. Where both upper and lower teeth are to be extracted the chair may be placed at a point intermediate between the high and low positions, or it may be in its high position and the operator avail himself of the use of stool while extracting the inferior ones. The chair, of course, must occupy its one position throughout the operation, as time will not admit of its being changed.

The operator should be careful never to begin operations until the patient is fully anæsthetized, nor should he continue after consciousness begins to return. Partial consciousness during any part of the operation will cause the patient to become alarmed and he may never again consent to being placed under anæsthetic influence. Ambition to extract a great many teeth under one administration has often led an operator to continue his operations after they should have ceased, and the result has been that the gas has been blamed for inefficiency where the operator alone was at fault.

It is always safer and better in every way not to attempt the extraction of too many teeth at one time. While our movements should be quick, they ought not to be hurried, for accident is more than likely to occur in the latter case. Extraction should never proceed too rapidly for the eye to follow every movement of the operator's hand.

In seizing a tooth the forceps should be fitted to the neck and then pushed well up under the free margin of the gum before pressure is made or removal attempted. In exerting the force for removal (after the tooth has been loosened in the socket by suitable movements) the operator should keep his arm as close to his body as possible, for he is thus able to apply more power than he otherwise could. It is also well to remember that in grasping the forceps the palm of the hand should never be turned upward, for wrist-power is materially lessened by such position.

Lancing the gums, except in the case of wisdom teeth, is usually unnecessary where enough of the crown remains to be properly grasped. Wisdom teeth, owing to the firm attachment of the soft tissues to their distal surfaces, should be lanced to prevent the laceration of these tissues. Serious difficulty has resulted in many cases where this has been disregarded.

Where roots are broken off or decayed below the gum line, and are firmly implanted, the gums should be lanced in order to enable them to be firmly grasped without injury to the soft tissues. The incision should be made on both the buccal and lingual surfaces in a vertical direction, extending about a quarter of an inch from the free margin. Lancing should, of course, be done before the administration of the anæsthetic is begun, and time should be allowed for entire cessation of the hemorrhage.

Some suggestions in reference to the order of extraction may perhaps be best formulated under the following rules:

1. Where both upper and lower teeth are to be

extracted, the lower ones should be removed first. If the order were reversed, the blood from the sockets in the upper jaw would obscure the view of the lower ones and interfere with their proper removal.

2. Where several posterior teeth in the same locality are to be removed, begin with the one farthest back and come forward.

3. Where teeth are to be extracted from both sides of the jaw, begin on the side having the greater number or presenting the greater difficulties. The importance of this will be apparent when we remember that the gag is always placed on the side having the least to be done, for the reason that we will have opportunity to do but little after it is removed.

4. Where both teeth and roots are to be extracted in different locations, it is usually preferable to remove the teeth first. By so doing we are often afforded better access to the roots. Sometimes it is better to reverse the order.

5. Where an entire jaw is to be cleared of teeth and roots, begin at the back on one side and come as far forward as time will permit. At the next sitting follow the same order on the opposite side. If some teeth in the front part of the mouth still remain, they may be removed at the third sitting. By dividing a large operation into different sittings we not only prevent interference from blood, but add greatly to the after comfort of the patient.

Forceps and Elevators.—While it is important

that the extracting instruments be sufficient in number and variety of form to cover all cases usually presenting, it is also most important to use as few of them at any one time as possible, since the change from one to another involves the loss of valuable time. Forceps with straight beaks and handles enable the operator to apply the greatest amount of force with the greatest safety. Where this form cannot be used, as in the lower jaw and posteriorly in the upper jaw, the curvatures or angles should be as slight as possible consistent with proper adaptation. A skilful operator will usually employ the same pair of forceps for the extraction of the ten anterior teeth and another for the six posterior ones in either jaw. While these four pairs, in addition to one or two pairs of root forceps, will be all that he will use in simple and favorable cases, he must have others in reserve for use under special circumstances. Among the latter there should be a Physick or a Stellwagen forceps for the removal of the lower wisdom teeth in difficult cases; also, forceps with very narrow beaks (curved for lower and straight for upper) for the removal of teeth crowded out of their normal position, or for roots where the crowns of adjoining teeth approximate so closely as to leave but little room for instrumental application. The forceps popularly known as the "cow-horn," for the extraction of the lower molar teeth, is an instrument of unusual power, but the danger of splitting the roots by its slight misapplication, or of allowing the tooth to escape from its grasp immediately after removal, is so great as to interdict its use during the anæsthetic state. The latter objection might apply with equal force to the use of the "Physick," but in some cases we cannot as well accomplish the result by the use of any other form. The use of forceps with spoon-shaped beaks, commonly known as "alveolar forceps," for the extraction of roots, cannot be too strongly condemned. By their use large sections are cut from the alveolar borders, leaving the ridge very irregular and making the future adjustment of plates quite difficult. Their use accomplishes nothing that may not be accomplished with the narrow-beaked root forceps.

Elevators, while their use is objected to by some operators, are held in great favor by others. The author has found them to be of such service in his practice that he would be very loth to relinquish their use. The principal objection to them is that they do not grasp the root, and therefore cannot control it after removal; but, used as they are, without previous lancing, in nine cases out of ten the root will simply be tipped over out of the socket and remain adherent to the ligament on the opposite side. The advantage of their use lies in the fact that with them we are often enabled to easily remove difficult roots, and sometimes whole teeth, without injury to either gum or process. Those illustrated in Fig. 13 represent the five forms found most useful in the author's practice. No. 1 is intended for the removal of individual roots anywhere in the upper jaw. After the point is forced well between the process and root, a rotary movement will easily dislodge the latter. No. 2 is used on roots of the lower jaw where we wish to push them inward toward the tongue. No. 3 is also intended for roots in the lower jaw, but is used to pull the root outward toward the lip or cheek. Nos. 4 and 5 are right and left instruments and prove serviceable in many ways, but especially where we wish to use an adjoining tooth as a fulcrum.

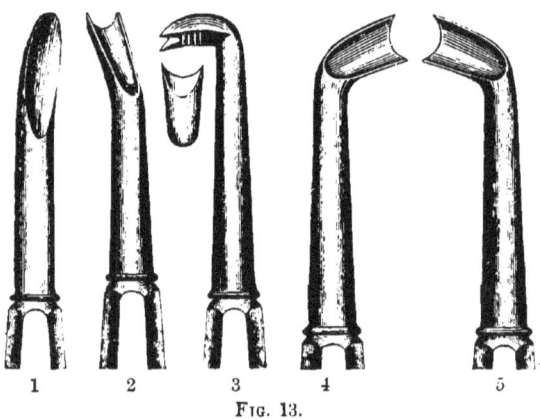

FIG. 13.

Silence being important during the process of anæsthetization, it is equally important during the period of recovery. The one condition is analogous to the other, and in either conversation is liable to confuse and alarm the partly bewildered patient. The gag, also, should not be removed from its position too soon after the operation is completed, for the patient is liable to construe it into an attempted removal of a tooth, and violent resistance on his part may follow.

As soon as the teeth have been removed the patient's head and body should be gently bent forward, so that the blood may run out into a hand spittoon provided for the purpose and not pass down into the stomach.

Should excessive hemorrhage follow extraction, it may usually be checked by inserting into the socket a tent of lint sprinkled with tannic acid. This may be kept in place, when necessary, by a compress of lint or linen placed upon it and then bandaging the head and jaw to prevent displacement.

Should this prove insufficient, pressure should be applied to the artery supplying the part at some point where it may be reached.

CHAPTER X.

ACCIDENTS AND EMERGENCIES.

Accidents may occur and emergencies certainly will arise at times in connection with the administration of any anæsthetic. To properly and promptly meet them requires that the practitioner should understand their variety, their characteristics. and their means of remedy.

Mr. Henry Sewill says: * "In the event of an accident the administrator ought to be able to recognize the symptoms and to act instantly. It is not enough to render the patient insensible and to trust in every case to an unaided recovery."

Dr. J. W. White has very tersely said: "When trouble comes to a patient from any cause during the anæsthetic state, it is not a good time to hunt up information."

Although accidents of any kind, associated with the administration of nitrous oxide have been remarkably few, they do at times occur, and no one knows when they may happen at his own hands.

---

*Medical Press and Circular. Reported in Cosmos, Vol. XV., p. 328.

The possibility of their occurrence should ever be kept in mind and the means of meeting them be always at hand.

Many accidents may be prevented by a preliminary examination of the patient and refusal to administer in all cases of a doubtful character.

Others, again (of a mechanical nature), may often be avoided by proper care on the part of the operator; while others still, such as paralysis of the heart or of the muscles of respiration, we have no means of either foreseeing or preventing.

Accidents of a minor character, or those least likely to prove fatal to life, usually occur through the lodgment of some foreign body in the larynx during the unconsciousness of the patient. Thus the broken beak of a forceps, the whole or part of a mouth-gag, or a tooth or portion of one, may easily be drawn into the trachea unless due care is exercised by the operator to prevent it. Cases of this kind have occurred, resulting, in some instances, in serious trouble, and in others causing death.

In 1867 a death occurred in the office of Dr. Lee,* in Philadelphia, during the administration of nitrous oxide. A post-mortem examination revealed the fact that an ordinary bottle-cork which had been used as a mouth-gag had passed into the trachea and become lodged there. Cases are not very infrequent in which a root or piece of tooth

---

*Cosmos, Vol. VIII., p. 384.

has been drawn into the trachea, causing sickness and trouble until finally dislodged in a fit of coughing.

Only a year or two ago the author was called into court as one of several expert witnesses to testify in a case of this kind. Gas had been administered to a patient by a competent practitioner for the extraction of a number of teeth. The operation was performed satisfactorily to the patient and without any knowledge of accident on the part of the operator. Shortly afterward the patient began to suffer from bronchial irritation attended with severe coughing and expectoration. The dentist was not advised of this, but the patient was treated by her physician, without any abatement of the troublesome symptoms. The patient became emaciated, and was finally obliged to take her bed under the impression that phthisis had set in. Some time later, in a violent fit of coughing, the root of a tooth was brought up, after which the patient soon regained her accustomed health.

Fortunately, the dentist was acquitted, it having been made evident to the jury that such an accident might occur to any practitioner, and was not, *per se*, an evidence of lack of skill.

A case very similar to this is recorded in the Dental Cosmos, Vol. XV., p. 478. Other cases of a like nature have been met with time and again that have not found record in the journals or gained the publicity of a law-suit.

ACCIDENTS AND EMERGENCIES. 73

To avoid all such, mouth-gags should be made of such substances as cannot be broken, and should be of such size as to be incapable of being drawn into the air-passages. A root or portion of a tooth may at times slip back into the fauces and be drawn into the trachea in spite of the greatest precaution, but every means should be adopted to prevent such an occurrence.

Once having occurred, if the trouble caused be not serious, we may wait a time for nature to expel the body ; but if the trouble be serious, the position of the obstacle should be located as nearly as possible and tracheotomy be performed.

An accident of minor importance, resulting in discomfort rather than danger, is luxation of the lower jaw. Sometimes, through the looseness of the condyloid articulation, and sometimes through excessive force applied in the extraction of a tooth, the jaw becomes dislocated. Should this occur, reduction is easily accomplished by placing a fulcrum between the teeth, as far back as possible, and then pressing the chin upward and backward. In this way the condyles of the jaw are released from their mal-position and caused to slip back into their proper place. A fulcrum may be extemporized by using one's thumbs, properly protected with a cloth or napkin, but a far better plan is to use, instead, two rubber bottle corks, which not only resist pressure moderately well, but act as rollers in assisting the backward movement of the jaw.

By far the most serious accident or emergency that may arise in connection with anæsthetization is asphyxia. Its approach will be indicated by increased and alarming lividity of the countenance and a struggle for breath. If continued, the breathing may become intermittent or cease entirely. It may occur through the slipping of some foreign body into the larynx; by the dropping backward of the tongue, thus preventing the ingress of air; or by temporary paralysis of the muscles governing respiration.

The first thing to be done is to draw forward the tongue with a tenaculum or artery forceps, or napkin. If this prove insufficient, the finger should be inserted to ascertain whether any foreign body has dropped into the fauces, and, if so, to remove it. Should both these operations prove ineffectual, artificial respiration should at once be resorted to.

In undertaking this, the patient should be lifted from the chair and laid upon the floor face upward. A pillow or other suitable article (the operator's coat folded or rolled will answer in an emergency) should be placed beneath the patient's shoulders and the head thrown well back.

The principle involved in artificial respiration is the aeration of the blood by artificial means. This is accomplished by causing the thorax to alternately expand and contract by manual aid, thus simulating the natural chest movements. When the chest is pressed downward and inward, whatever air there may be remaining in the lungs is forced out, and when by a reverse movement the thorax is expanded a vacuum is produced which will cause the air to rush in and occupy the space.

There are several methods of accomplishing artificial respiration, but the following are among the simplest and best:

1. *Simple Manual Pressure.* With the patient in position as described, the operator should place the palms of both hands on the ribs close to the sternum and near its lower end and force them well down to empty the lungs. By releasing the pressure the natural resilience of the parts will cause them to resume their former position, thus inducing inspiration. These alternate movements should be repeated rapidly until the muscles resume their normal function and breathing becomes automatic. Mr. Harley says: * " Manual pressure equal to about thirty pounds may be with perfect safety applied to a healthy adult human thorax." Again: "The manual pressure ought to be made on the lower part of the sternum, for the resilience of the thoracic walls is there greatest; and pressure on the abdomen at the same time is not to be omitted, or the diaphragm will descend and counteract the benefits derived from the pressure made on the lower part of the chest."

2. *Howard's Method.* This is more forcible and

---

*Holmes' System of Surgery, Vol. 3, pp. 104-5.

efficient than the method just described. It consists in producing contraction of the thorax by means of the operator's knee, while expansion is produced by pressing the patient's arms upward and backward, alternating the two movements as before. The position of patient and operator, as well as the *modus operandi* are well illustrated in Fig. 14. Both

Fig. 14.

of the methods just described may easily be carried out by the operator alone, but where he has the assistance of others the following plan may be adopted :

3. *Sylvester's Method.* This is considered the best method of producing artificial respiration and is the one most generally resorted to. In it both expansion and contraction of the thorax are produced by movements of the patient's arms, accompanied by suitable pressure. Mr. Harley says (loc.

cit.) : "On bringing down the patient's arms they should be gently and firmly pressed against the sides of the chest, so as still farther to diminish the cavity of the thorax. This pressure can be exercised with greater facility and equal effect by pressing the arms on the lower third of the sternum. By alternating the movements of the arms and

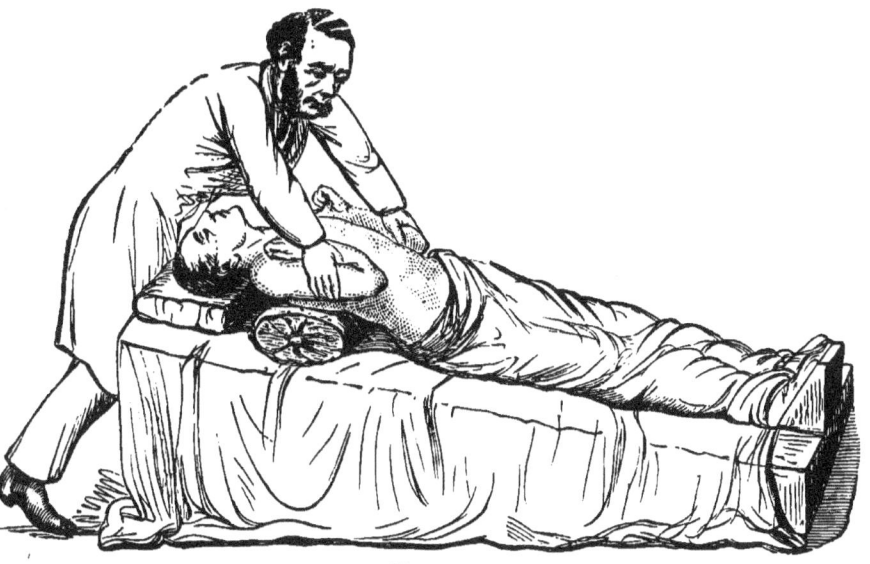

Fig. 15.

pressure of this kind a regular exchange of air can be produced, varying in quantity from thirty to fifty cubic inches, an amount more than is requisite for the purposes of resuscitation.

In all cases the respirations should amount to at least thirty or even forty per minute. The natural respirations are only eighteen per minute, but in

78   NITROUS OXIDE.

cases of resuscitation, as our object is to arterialise the blood even more rapidly than in health, and as we cannot introduce by artificial means the same amount of air that is taken in by the normal efforts, we must proportionally increase the number of respirations."

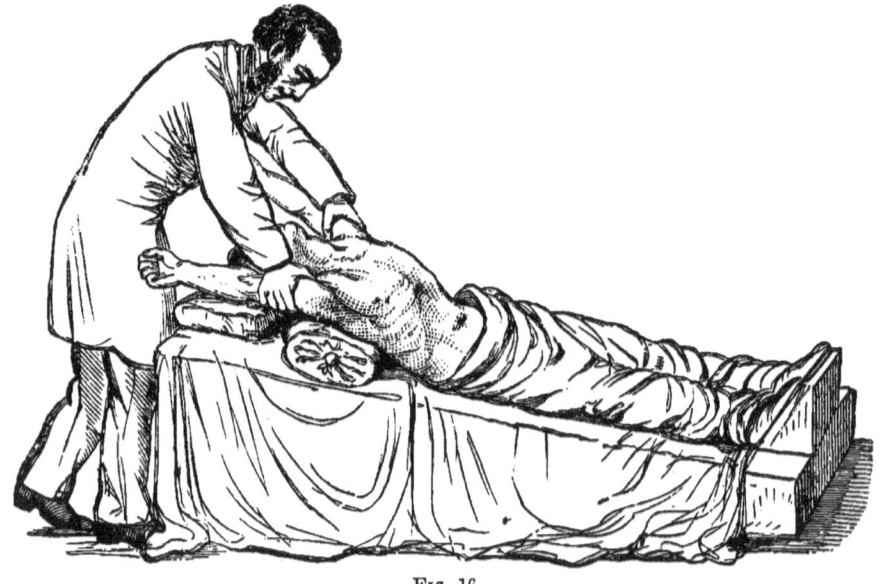

Fig. 16.

Figs. 15 and 16, copied from Holmes' Surgery, illustrate perfectly the "Sylvester method."

CHAPTER XI.
COMBINED ANÆSTHETICS.

The combination of different anæsthetic agents in order to secure the advantages of each without the disadvantages of either has long been a subject for experiment. In medical practice various combinations of ether, chloroform and alcohol have been tried for the past quarter of a century or more with results that have been approved by some and condemned by others. Ether and chloroform were combined in order that the safety of ether and rapid action of chloroform might be conjoined, while chloroform and alcohol were united so that the depressing effects of the former might be counteracted by the stimulating effects of the latter. Each of the various combinations has its advocates and opponents, so that medical sentiment to-day is divided in regard to the advisability or otherwise of combination. It is believed, however, that combination in one form or another is growing in favor, and that there are quite as many practitioners who use a mixture as there are those who use a single agent.

The efforts to combine nitrous oxide with other agents are of rather recent date. The first experiments in this direction were probably those of M. Paul Bert, begun in 1879, in which by combining nitrous oxide with pure oxygen he hoped to prolong the anæsthetic state and render it suitable for major operations. His experiments, however, with the combination under ordinary conditions, as well as the exhibition of it under increased atmospheric pressure, resulted in nothing of practical value.

The next attempt to use nitrous oxide in connection with another agent was made in England. In order to overcome or avoid the pungent effect of ether vapor upon the glottis, it was suggested that the anæsthetization should be begun and carried forward for a time with nitrous oxide, after which ether should be substituted and unconsciousness maintained. To facilitate this operation a special inhaler was devised, which seemed to meet with much favor.

Appreciating the value and relative safety of nitrous oxide as an anæsthetic, and at the same time recognizing the limitation of the anæsthetic state produced by it, various attempts have been made in this country during the past ten years to combine it with some other agent by which its effects might be prolonged and its safety not interfered with.

One of the first, if not the first, combinations of nitrous oxide with another agent was introduced to the profession many years ago under the name of "Mayo's Vapor." Being a proprietary article, the character of the compound was not made known.

It was placed upon the market in the ordinary iron gas cylinder, ready for use, and when drawn from the cylinder the gas or vapor was made to pass through a wash-bottle containing a solution of wintergreen either to conceal the real odor of the compound or to make it more pleasant for the patient to inhale. Notwithstanding the secret character of the compound, which should have deterred a professional man from employing it, it found a ready sale and is to-day quite extensively used in some of the Eastern and Middle States. So far as we have been able to ascertain, it is the only article of an anæsthetic character in which the different constituents are combined and compressed in a gas cylinder.

Usually, where nitrous oxide is used in combination with another agent, the pure liquified gas in its escape from the cylinder is made to combine with the vapor of another anæsthetic agent (usually ether or chloroform) in a receptacle intermediate between the cylinder and gasometer.

Fig. 17 represents an attachment for gas cylinders, devised by Dr. A. M. Long, in which the gas is made to combine with the vapor of some liquid anæsthetic. C is the combining chamber through which the gas passes from the cylinder in the direction of the arrow on its way to the gasometer. A is the receptacle for the liquid agent. When the handle below it is turned half-way around, the two separated tubes meet at B, and one drop of the liquid passes down into the combining chamber. As each turn permits the passage of a drop only, the administrator knows with exactness the amount he is introducing into the chamber. An appliance such as this, or one similar in principle, is generally used by those who make their combination after this manner.

As to the advisability of combining nitrous oxide with some other more potent agent, opinion is

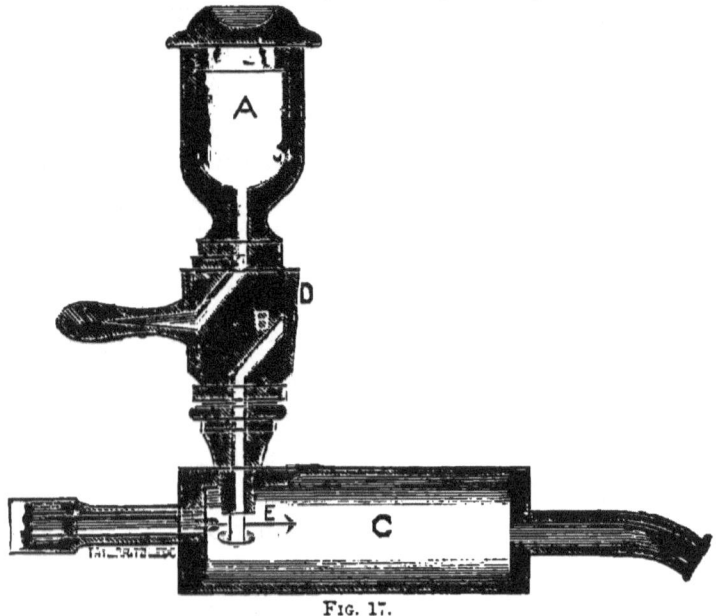

Fig. 17.

divided. The bulk of sentiment in the dental profession to-day is opposed to it on the ground that by introducing a more powerful agent (and necessarily a more dangerous one) we step aside from ground that is relatively safe to that which is less so, and in so doing we imperil the lives of our patients to that extent. It is also argued that each anæsthetic agent possesses properties peculiar to itself and operates in a manner different from any other agent, and that by combining them, in case any unfavorable symptoms appear in the process of anæsthetization, we are at a loss to know how to apply restorative measures, such as we could safely and intelligently apply where but one agent is used.

Those who favor combinations argue that the amount of vapor of the more powerful agent is so small in comparison with the quantity of nitrous oxide with which it is combined that the safety of the gas is not impaired by it. They also cite the fact that, notwithstanding the many years during which such combinations have been in use, no accident or unfavorable result has been known to follow their exhibition either in dental or medical practice.

With a view to ascertaining the record that the combination of gas with ether or chloroform had made for itself, the author addressed inquiries to several dentists who have had large experience in that method of practice, and in each case has received replies most favorable in character.

One gentleman writes : "I always use chloroform and alcohol, equal parts, in combination with nitrous oxide. From three to six drops of the mixture with five gallons of gas I find to be about the proper proportion. I cannot see any difference between the effects of the combination and those of simple nitrous oxide except that the former lasts a little longer. The recovery in each case seems to be equally rapid. I have had no unfavorable or unpleasant effects following the administration of the combination, and believe it to be quite as safe as nitrous oxide alone."

Another gentleman says: "I use equal parts of the best chloroform and absolute alcohol, say from one to two drops of the mixture to each gallon of gas. By its use I obtain quicker anæsthetization, while the recovery is a little slower than when gas alone is administered. The pulse seems to be firmer and stronger under the combination, and there is less discoloration of the face. I have had cases where the nitrous oxide would not thoroughly anæsthetize, but no case in which the combination failed. For five years I used simple nitrous oxide, but for the past six years have used the above combination and greatly prefer it. Have administered it many hundreds of times without a single unfavorable result."

Still another writes: "I have used nitrous oxide for eighteen years, and during the past five years have given it in combination with either chloroform or ether. For three years I used chloroform with the gas, but for the last two have used Squibb's ether instead, and prefer it. I combine from fifteen to twenty-five drops of the ether with five gallons of the gas. By using the combination I can extract as many as sixteen teeth and roots at one time, whereas with the gas I have seldom been able to remove more than five or seven. Patients seem to recover from the effects of the combination as quickly as from gas. I have had no nausea or other unpleasant effects to follow its administration. I cannot suppose a case where it would be proper to give the gas in which I should hesitate to administer the combination. During the past five years I have used from eight to twelve hundred gallons of gas each year in combination with some other agent, and having had no ill results in all that time, must consider it quite as safe as any anæsthetic agent can be."

Coleman, speaking of the combination of ether with nitrous oxide (the latter given alone up to a certain point, followed by the former), says: *
"Ether, although no doubt a much safer anæsthetic than chloroform, is yet probably less safe than nitrous oxide. Owing to its pungent and irritating nature, it is less agreeable to inhale than either nitrous oxide or chloroform, but this may be overcome by employing it as Clover was the first to do, viz., in combination with nitrous oxide. Ether with nitrous oxide is most suitable for cases in which several teeth are to be removed at one sitting. We think it also well calculated for those in which severe after-pain may be anticipated, as in a tooth giving rise to alveolar abscess in any of its

---

*Loc. Cit., p. 339.

stages, recovery being less sudden than with nitrous oxide alone. With regard to symptoms indicating danger, these will be much the same as with nitrous oxide alone, though perhaps less apparent, and consequently requiring greater watchfulness, as with ether there is usually less lividity. Whilst the respiration is carefully watched, the pulse must be more regarded than with nitrous oxide. In employing the ether combination we must look for more tendency to vomiting."

The author has had no experience with the combination of nitrous oxide and ether except in an experimental way, and therefore is not prepared either to approve of or condemn it, but, in view of the favorable record it has made for itself in the past six years in this country, he is inclined to look upon it with favor, and believes that further experimentation with it will result in a general acknowledgment of its value in cases where an effect intermediate between that of nitrous oxide and of ether is desired.

CHAPTER XII.

LEGAL CONSIDERATIONS.

While the professional man is ever liable to be held legally responsible for any accident or mischance that may occur in his practice, he is naturally held to a greater responsibility when any misfortune occurs to a patient under anæsthetic influence. In the one case it happens while the patient is in possession of all his faculties, and the fault, if there be any, may be partly his own; in the other the patient is unconscious, and the whole responsibility is necessarily thrown upon the operator. Intelligent and right-minded people will not be inclined to blame or bring trouble upon an operator for any ordinary accident occurring in the line of his duty, but the ignorant and those evilly disposed are only too ready to resort to the law to obtain compensation for real or fancied injuries.

In view of this, the practitioner should familiarize himself with the legal responsibilities under which he, as a specialist, is placed.* The law,

---

*The author would especially recommend to all dentists and physicians a work entitled "Williams on Laws Relating to Physicians, Dentists and Druggists," by R. J. Williams, of the Philadelphia Bar.

while guarding the interests of the patient, is equally watchful of the rights of the practitioner. It requires of him that he shall be possessed of ordinary skill, and that in the employment of it he exercise ordinary care and diligence. It also demands that he shall be of good moral character.

In the use of an anæsthetic, it would demand, in addition, that he be familiar with the character of the agent used, its method of action, complications that may be associated with its use, and the remedial measures with which to meet them.

In case of death occurring under the influence of an anæsthetic, the operator, in order to be free from blame in the eyes of the public, as well as the law, would have to show that all the above requirements were met, and that all means usually employed to prevent such a catastrophe had been employed.

When this can be done, no judge would charge or jury decide that the practitioner was responsible for the misfortune. Should the practitioner, however, not be able to show that he had exercised "ordinary care, diligence and skill," he might be held liable for heavy pecuniary damages, or, possibly, be found guilty of manslaughter.

Happily, few deaths have occurred in connection with the administration of nitrous oxide, and none of them very recently; but, with all this, it is most important that the operator should realize in each case the possibility of death occurring, be on the

LEGAL CONSIDERATIONS.

alert for any unfavorable symptoms, and have at hand all means for immediate attempt at resuscitation.

In case of death resulting in connection with the administration of an anæsthetic of a proprietary or secret character, the operator could hardly hope to escape conviction in a case at law, for it would be held, and justly so, that in administering a compound the character and ingredients of which he is ignorant of, any attempt at counteracting the unfavorable symptoms would necessarily have to be conducted blindly, and, therefore, unskilfully.

Scarcely less serious than the charge of manslaughter is that of criminal assault. This charge has so often been preferred by females against the administrator of an anæsthetic that it becomes all important to protect one's self against it. The charge may be made maliciously for the purpose of blackmail, or it may be made in all honesty by one who believes that she has been wronged while in an unconscious condition. In most cases (we hope in all) it is a delusion caused either by a dream or by that confusion of ideas and emotions likely to occur during the abnormal state of anæsthesia. In medical practice, it is a generally admitted fact that anæsthetics do stimulate the sexual function. To this rule nitrous oxide is no exception, as abundant evidence proves. It is also admitted that this sexual stimulation is more frequently induced (under an anæsthetic) in females than in males, especially at the time of periodical pelvic congestion. These facts, taken in connection with the mental disturbance before alluded to, and the knowledge of the patient before passing under the influence that she will be entirely at the mercy of the operator, cause it scarcely to be wondered at that delusions of this character should occur. One writer truly says: " Cases have occurred in which the woman was so positive that liberties had been taken with her person during anæsthesia that the testimony of relatives who were present all the time scarcely sufficed to convince her that she was laboring under a delusion."

Dr. B. W. Richardson, of London, has related a case where the patient, a female, was being operated upon by a dentist, and alleged that he had criminally assaulted her; and this statement she persisted in, though her own father and mother, Dr. Richardson, and the dentist's assistant were all present during the entire time.

Another case is on record where a woman made a similar charge, although her husband was present during the whole period of anæsthesia.

Wharton & Stille * relate a case in which, while enlarged nymphæ were being removed, the woman unconsciously went through the movements of sexual organism in the presence of numerous bystanders.

The case of Dr. Beale, the Philadelphia dentist, is still fresh in the minds of some of the older prac-

---

*Medical Jurisprudence.

titioners. After the administration of ether to an unmarried woman for the extraction of teeth, he was accused by her of criminal assault. Undoubtedly innocent, he was nevertheless convicted of the crime upon the unsupported testimony of the woman herself, and sentenced to ten years' imprisonment, though, fortunately, afterward pardoned. Strange to say, a physical examination of her person at the time or before the trial was not made, and thus one of the best evidences of guilt seemed to have been entirely overlooked. Fortunately, with our present increased knowledge of the action of anæsthetic agents, and of the peculiarities attending their administration, a similar verdict under similar circumstances could scarcely be reached at the present day.

The dentist who confines himself to the use of nitrous oxide, and avoids the more powerful agents, is not likely to be accused of criminal assault, as the duration of the anæsthetic influence is too short to admit of such a result. It will not exempt him, however, from the possible charge of attempted liberties, and to avoid this, as well as charges of any other character, it becomes his duty to protect himself by always having a third person present when he administers an anæsthetic to a female.

Legal complications may follow accidents of any nature occurring at the dentist's hands. Those of a minor character, such as luxation of the inferior maxilla, injury to the soft tissues, or the lodgment of a root or foreign body in the trachea during extraction under an anæsthetic may be the means of causing him much trouble should the patient decide to appeal to the courts for indemnity for the injury inflicted.

Our only means of protection against such possibilities is to use our best care and skill to prevent accidents; and should they occur, in spite of us, we should be able to convince both judge and jury that "ordinary care, diligence and skill" had been exercised. Being successful in this, we need not fear the result of the verdict.

One more consideration, having a legal bearing, remains to be noticed, namely, antisepsis. In the present light of medical science, the induction of disease by transmission has received, and is receiving, the attention which its importance demands. Although the germ theory of disease has not been fully accepted by a large number of medical practitioners, it has been accepted by so many of high standing, and has taken such a firm hold upon the general public, that scarcely a surgeon would venture to operate without the aid of antiseptic treatment in some of its forms. Indeed, a surgeon would hardly feel safe from the terror of the law did he not employ some means of preventing the transmission of disease through the medium of his instruments. If this be important for the general surgeon it must be equally so for the dental surgeon, for the same conditions exist in both cases.

LEGAL CONSIDERATIONS. 93

Cases have occasionally occurred in which the induction of a disease has been attributed (justly or unjustly) to virus transmitted by means of the dentist's instruments. Everyone must recognize the possibility of such an occurrence, although the probability of it would naturally be very slight. Uncleanliness or lack of care in the treatment of instruments before or after use would be as likely to produce ill results at the hands of the dentist as of the surgeon. The virus of syphilis, for instance, might be conveyed by means of extracting instruments from the mucous patch in one mouth to the healthy but lacerated gum in another.

Now and again physicians, in searching for the cause of syphilitic lesions in their patients, after eliminating all other probable causes, have come to the conclusion that the virus must have been conveyed by the dentist's instruments in the extraction of a tooth.

The author is familiar with a case of this nature that occurred in one of the New England States some years ago. A lady, the wife of a physician, went to her dentist and had a tooth extracted. Some time afterward symptoms of syphilis began to manifest themselves in the falling of the hair, mucous patches in the mouth, and well marked syphilides on parts of the body. The physician, finding no other cause to which to attribute the infection, concluded that she must have been inoculated by means of the dentist's forceps. He so informed the dentist, who, while he doubted the probability, could not deny the possibility, and so the matter rested. No legal action was taken, but the physician and his wife removed to a distant portion of the country to hide the seeming disgrace and begin life anew among strangers.

Whether the dentist was really at fault in the matter no one can tell, but all such instances emphasize the importance of thorough cleanliness in the care of dental instruments, and the disinfection, by means of mercuric bichloride solution, or other germicide, of all such instruments and appliances, as necessarily come in contact with open wounds in the mouth.

www.ingramcontent.com/pod-product-compliance
Lightning Source LLC
Chambersburg PA
CBHW020157170426
43199CB00010B/1074